PANORAMA

A HISTORY *OF*

INVENTIONS

FROM ABACUS *TO* ATOMIC POWER

Series designer: David Salariya
Editor: Penny Clarke
Artists: David Antram
Ray and Corinne Burrows
Nick Hewetson
Mark Peppé

Illustrations by:
David Antram 36-37, 38-39, 40-41, 42-43;
Ray and Corinne Burrows 24-25, 26-27, 28-29,
30-31; **Nick Hewetson** 14-15, 16-17, 18-19, 22-23;
Mark Peppé 8-9, 10-11, 12-13, 20-21, 32-33, 34-35.

First published in 1997
by Macdonald Young Books,
an imprint of Wayland Publishers Ltd.
61 Western Road
Hove
East Sussex
BN3 1JD

ISBN 0-7500-1875-5

Printed in Hong Kong by Paramount Publishing Co. Ltd.

A CIP catalogue record of this book is available from the
British Library.

Author:
Peter Lafferty is a former secondary school
science teacher. Since 1985 he has been a
full-time author of science and technology
books for children and family audiences.
He has written 50 books and contributed
to many well-known encyclopedias. He
has also edited scientific encyclopedias
and dictionaries. He lives in Sussex and
when not writing books or walking his
dog, he can usually be found fishing
for trout.

Series designer:
David Salariya was born in Dundee,
Scotland, where he studied illustration
and printmaking, concentrating on book
design in his post-graduate year. He later
completed a further post-graduate course
in art education at Sussex University. He
has illustrated a wide range of books on
botanical, historical and mythical
subjects. He has designed and created
many new series of children's books for
publishers in the UK and overseas,
including the award-winning **Inside Story**
series. He lives in Brighton with his wife,
the illustrator Shirley Willis.

PANORAMA

A HISTORY OF

INVENTIONS

FROM ABACUS TO ATOMIC POWER

Written by
PETER LAFFERTY

Created & Designed by
DAVID SALARIYA

MACDONALD YOUNG BOOKS

Contents

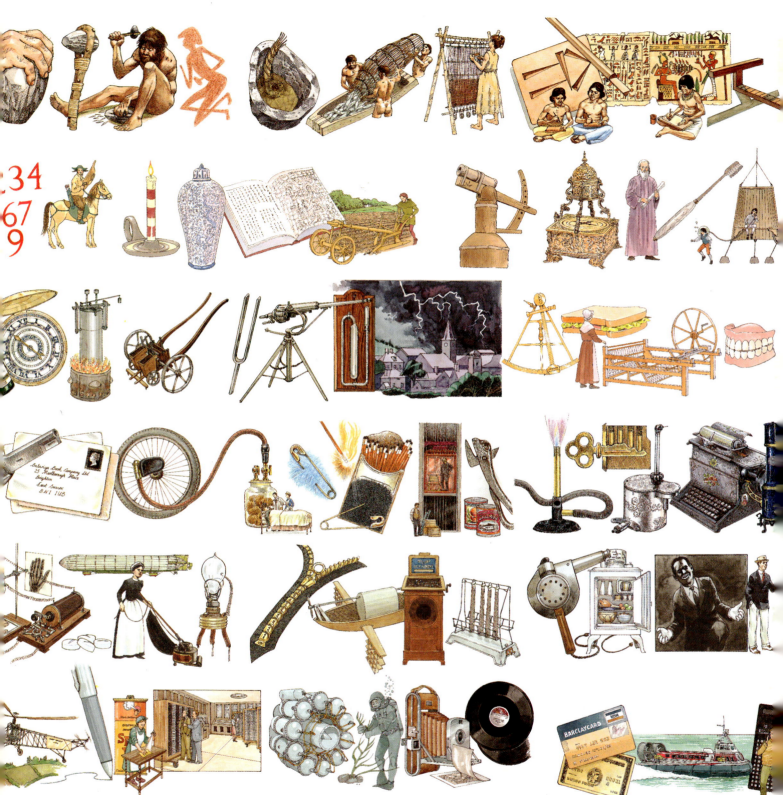

INTRODUCTION

Look around your home. You will see many amazing inventions. Perhaps there is a radio, a television set, a CD player, a video recorder, an electric cooker and printed books and newspapers. These inventions make life today much more pleasant than it was in the past.

Many of the inventions which give us an easier way of life have been produced in modern times. The radio, television and computer are 20th-century inventions, for example. However, inventors have been at work since ancient times.

The cave-dweller who first sharpened a rock, so producing the first axe, was an inventor. So was the person who made the first wheel – an ancient invention that is still vital today.

Inventors built on these simple beginnings. Often one invention leads to another. For example, the invention of the electric battery led to the invention of the electric motor. Over the years, the pace of discovery quickened. Each new scientific discovery was quickly used to create new gadgets and machines to make life easier and more fun. This book tells how we progressed from the stone axe to the computer. It is the story of inventors and their inventions.

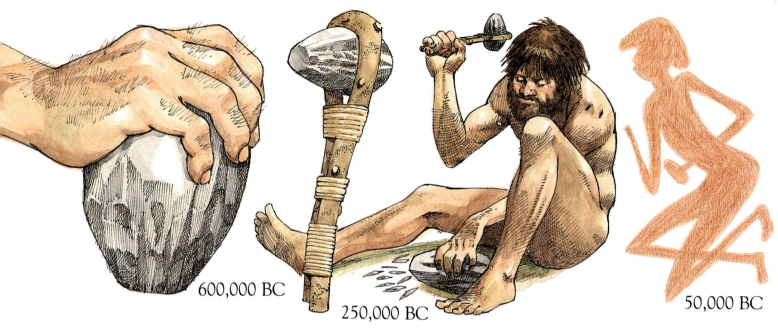

600,000 BC

250,000 BC

50,000 BC

THE FIRST INVENTIONS

An Egyptian wooden sickle set with flint teeth, used to harvest grain around 6000 BC.

Many early inventions were probably due to luck. Perhaps, around 600,000 years ago, a cave-dweller used a stone to crack a nut. Then the stone chipped, giving it a sharp edge that could be used to cut meat. The stone had become a simple knife!

Other early inventions were the result of careful thought. The bow and arrow, which was invented in Africa about 30,000 BC, could not have been invented by accident. The first known looms for weaving cloth were complicated devices – they needed thought and skill to make. Our early ancestors also discovered how to work metals, make pottery and baskets, and build strong houses.

By around 3500 BC, life in some parts of the world was much more comfortable. In Mesopotamia (now part of Iraq) people lived in villages with houses made of mud bricks. They grew food crops in fields around the village. Tools and weapons were made from metals such as bronze. Pottery was made on a potter's wheel, just as it is today.

ABOUT 7000 BC, simple pots were shaped from coils of wet clay (above). The potter's wheel (below) was invented around 3500 BC in Mesopotamia.

BRONZE is a mixture of copper and tin which came into use around 5,000 years ago. Molten bronze was poured into a mould and allowed to cool.

CLAY, a type of mud, was one of the basic building materials of the ancient world. It was dug from moist river beds.

BY 3000 BC, the Egyptians were mixing chopped straw with clay to give extra strength to the bricks.

50,000 BC

8000 BC

7000 BC

THE oldest known bricks, more than 8,000 years old, were found in the walls of Jericho in Jordan.

BY ABOUT 3500 BC, farmers in Mesopotamia were harnessing oxen to ploughs to break up the soil before planting crops.

THE FIRST rectangular houses were built in the valley of the River Jordan around 7000 BC. Permanent houses had been built 2,000 years earlier but these were circular.

BRICKS were hardened in the sun. In Mesopotamia about 3,500 years ago, the kiln or oven was invented to dry bricks more quickly.

600,000 BC A stone used for cutting – one of the earliest inventions. The earliest stone tools were shaped to a point or cutting edge at one end and rounded at the other end to hold.

250,000 BC Stone axes were used in Europe, Asia and Africa. The axe head was bound to a wooden handle by leather strips. Adding a handle meant the axe head could be swung with greater force than a hand-held axe.

50,000 BC Ancient cave paintings have been found in the Middle East, Europe and Africa. The paints were made from coloured muds and rocks. Many paintings show hunting scenes.

50,000 BC Cave paintings would have been impossible without some form of lamp. These were made of hollowed-out stones which held a lump of animal fat with a wick made of moss. The oldest known lamp is 17,000 years old and was found in France. However, burn marks on cave walls show that lamps were used over 50,000 years ago.

8000 BC Fish traps meant bigger catches of fish than were possible with lines or spears. The traps, woven from thin wooden rods and twigs, were placed across the mouths of streams. Traps like this are still used for catching salmon in Sweden.

7000 BC The earliest loom for weaving cloth consisted of two wooden rods pegged to the ground. Threads were stretched between the rods while other threads passed between them at right angles. The loom shown dates from 1500 BC.

3000 BC

3000 BC

2000 BC

THE WHEEL

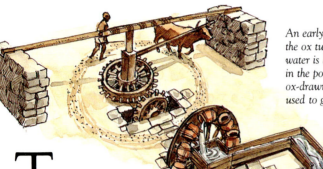

An early water-wheel. As the ox turns the main wheel, water is lifted from the well in the pots. Similar ox-drawn wheels were used to grind corn.

1   2

3   4

Early wheels:
1 solid, made of wood or stone;
2 plank wheel, made of planks;
3 semi-solid, with small cut-out sections;
4 cross-bar wheel, with spokes for strength.

BEARINGS, which make wheels easier to turn, appeared around 100 BC. They consisted of wooden rollers around the axle.

T he wheel is one of the most important mechanical inventions of all time. It is not just used on road vehicles as a means of transport, but also in the potter's wheel, and in gear wheels and pulleys.

Before the wheel was invented, heavy loads were dragged along on sleds or rollers made of tree trunks. The rollers had the same effect as wheels but a lot of work was needed to keep the rollers in place as the load moved along.

The first wheel appeared in Mesopotamia around 3500 BC. This was the potter's wheel – a circular table which could spin while a potter worked clay on it. Three centuries later, simple vehicle wheels were used in Mesopotamia. Early wheels were sometimes solid discs of wood cut from tree trunks. Another early wheel design used planks of wood fastened together with wooden or metal brackets. These wheels were heavy, so sometimes sections of the wood were cut out to make them lighter. It was then a small step to the spoked wheel, which appeared in Mesopotamia around 2000 BC.

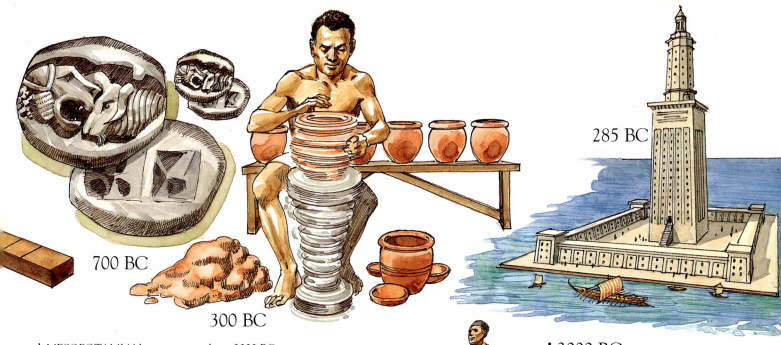

700 BC

300 BC

285 BC

MESOPOTAMIAN street scene about 2000 BC. Several types of wheeled vehicle can be seen. The cart (bottom left) has wheels made from planks clamped together; the other cart (bottom right) has cut-out wheels strengthened with struts. The spoked chariot wheels are lighter than plank wheels and helped make the chariot into a high-speed war machine. The wheel rim was made by bending a single piece of wood into a circle or by using several sections joined together.

A CHINESE wheel barrow invented in the 1st century BC. It needed less effort to move because the load was balanced on either side of the wheel.

3000 BC An early form of writing known as cuneiform was used by the Persians and other Middle Eastern peoples. Cuneiform used simplified pictures for words and ideas. The symbol for 'bird' is shown. The symbols were formed by pressing the surface of a damp clay tablet with a piece of reed that was triangular at one end.

3000 BC The Egyptians introduced hieroglyphics, a form of picture writing, to represent words, syllables and sounds. When papyrus, an early form of paper, was invented, the symbols had to be changed so that they could be written more rapidly. This was the beginning of our system of writing.

2000 BC An Egyptian shadow clock. The shadow of the cross-bar fell on a scale which indicated the time. In the morning, the clock was pointed east into the sun. In the afternoon, the clock was pointed westward.

700 BC King Gyges of Lydia (in present-day Turkey) is thought to have issued the first coins. Made from a mixture of gold and silver, they had a picture of the king stamped on one side.

300 BC The potter's kick wheel was used by the Greeks and Egyptians around 300 BC. The potter kept the pot turning by kicking on a heavy wheel at the bottom.

285 BC The first lighthouse was called the Pharos. It was built off Alexandria, in Egypt, around 285 BC by King Ptolemy II. The lighthouse was over 130 metres high. A fire was kept burning all night at the top.

200 BC

150 BC

150 BC

ARCHIMEDES' MACHINES

Archimedes lived from about 287 BC to 212 BC in the Greek city of Syracuse in Sicily. He was a famous mathematician and astronomer as well as an inventor.

Archimedes was the greatest inventor in ancient Greece. He was the first person to make a scientific study of how levers, screws and pulleys worked. Using his discoveries, Archimedes built machines capable of lifting huge weights. Once King Hieron of Syracuse set a test for Archimedes. Pointing to a large ship that lay ashore near the harbour, he ordered Archimedes to move the ship single-handedly. Archimedes made a large pulley and used it to drag the ship into the water.

When Syracuse was attacked by the Romans in 215 BC, Archimedes used his inventions to defend the city. He is said to have placed huge curved mirrors on top of the city walls, and focussed them on the Roman ships, setting them alight. If any ship came near the city, it was grabbed by iron hooks and lifted from the water using huge levers, and dropped back into the water from a great height. The Romans withdrew and decided to starve the city into submission instead. The siege lasted three years. When the Romans eventually entered the city, Archimedes was killed by a soldier as he worked on a mathematical problem.

BEFORE Archimedes' time, water was lifted from low-lying rivers and channels using a shaduf. This consisted of a long wooden pole which acted as a lever, with a weight on one end. The other end held a bucket which was lowered into the river and filled with water before being lifted.

A PULLEY is a machine with a rope which runs over one or more grooved wheels. When the rope is pulled, the pulley increases the pulling force. Archimedes used pulleys to move a loaded ship without help.

ARCHIMEDES invented a screw (now called the Archimedean screw) which lifted water from canals and rivers onto fields. These screws are still used in Egypt today.

150 BC

150 BC

10 BC

THE ABACUS, possibly the earliest calculating machine, was invented by the Romans. It remained popular in the Far East for many centuries after it was no longer used in Europe.

THE Archimedean screw consists of a hollow tube containing a spiral that can be turned by a handle at one end. When the lower end of the tube is placed into water and the handle turned, the water is carried up the tube and splashes out onto the ground. The screw is a simple machine – it reduces the effort needed to lift the water.

200 BC The use of the arch in buildings and bridges was an important development. Before the arch was invented buildings were built with vertical pillars supporting horizontal beams. The Romans first built arch bridges around 200 BC.

150 BC The first true paper was made in China around 150 BC. Cloth, wood and straw were beaten to a pulp and mixed with water. Thin sheets were pressed out and hung out to dry.

150 BC The Romans were the first to develop central heating. Heat from a furnace was fed under the floors of houses through channels in the brick walls. The idea was lost when the Roman Empire collapsed in AD 410.

150 BC Greek inventor Ctesibius of Alexandria, Egypt, made a suction pump. It used air pressure to suck water up as the handle at the top was rocked back and forth. Ctesibius also invented a water clock and made the world's first organ.

150 BC The screw press described by the historian Pliny of Rome. The press was used to press oil from olives and juice from grapes. The handles were turned to press a board down on the fruit.

10 BC The Roman engineer Vitruvius described the inventions of Archimedes in a handbook he wrote for architects. He described a treadmill-powered crane which consisted of a pole with a pulley at the top. The weight was lifted by turning the wheel or treadmill at the bottom. Cranes of this type were used until the 1800s.

cAD 300

1234
567
89
c500

c850

c850

EARLY WEAPONS

Armour through the ages:
Roman (left),
Norman (middle) and
steel plate (right).

The first weapons were clubs, spears and stone axes. Over time, weapons became more effective and could kill over greater distances. The bow and arrow was invented about 30,000 years ago. It enabled people to kill beyond throwing range for the first time.

The first armour, made of small bronze scales sewn on to a leather backing, was used in Mesopotamia around 2000 BC. The Romans used armour made of iron strips wrapped around the body in 100 BC. In AD 1000, the Normans wore chain-mail armour, made up of small linked metal rings. In the 14th century, steel plate armour covering all parts of the body was made in Europe.

Gunpowder was invented in China around AD 950, reaching Europe around 1242. English scientist Roger Bacon described how to make it in 1249. Gunpowder changed the nature of warfare. The catapult and battering ram were replaced by the cannon. Small handguns replaced the crossbow and longbow. Now, castle walls could be smashed down by cannons and armour could be pierced by bullets from handguns.

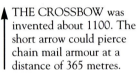

THE CROSSBOW was invented about 1100. The short arrow could pierce chain mail armour at a distance of 365 metres.

THE FIRST recorded use of cannon in Europe was in 1346 when Edward II of England defeated the French at the Battle of Crecy, France, although the English longbowmen probably contributed more to the English victory than the cannons. The cannons were more successful when used against the walls of a castle or town (right). At first, cannons fired stone shot; iron cannon-balls were not introduced until about 1400.

The powerful longbow was invented around 1250 in Wales. It could be fired more quickly than the crossbow.

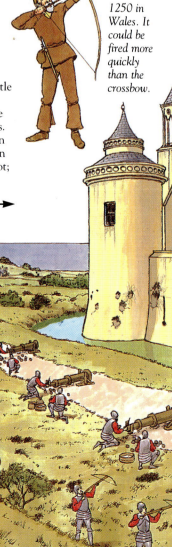

868 c950 c1320

AROUND 500 BC, the catapult was used by the Greeks to bombard cities with large rocks.

A muzzle-loading cannon of 1854. A bag of gunpowder and the shell were pushed down the barrel from the front or muzzle.

cAD 300 The stirrup was invented in China around 300. A century or so earlier the Chinese had also invented the padded saddle. Stirrups and a saddle gave the rider a firm seat on the horse.

c500 The numbers we use today were invented by the Arabs. The value of a digit depends upon its position in a number.

c850 King Alfred the Great of England is said to have invented the candle clock. A candle was marked with hours. As it burned down, the hours passed could be read from the marks.

c850 The Chinese first made vases and cups of porcelain, a tough white material made from clay. It took another 350 or so years before the new material found its way to Europe.

868 The first printed book, the *Diamond Sutra*, was made in China. It was printed with carved wooden blocks and showed scenes from the life of the Buddha, a religious leader.

c950 The wheeled plough was invented in Europe. It had a pair of wheels which made it easier to pull and control.

c1320 The 'pot de fer' was an early form of cannon invented in France. Arrows were fired from the barrel using gunpowder.

1400   1452   1498   1538

THE PRINTING REVOLUTION

A page from the Bible printed by Johannes Gutenberg in 1455. Two hundred copies of it were printed.

A PIECE of metal type with the raised letter at one end. The pieces were fitted together to form words.

A compositor inking type with a dubber.

Printing is one of the most important inventions of all time. Before the 15th century, all books were written out by hand using quill pens. It took months, or even years, to make a single copy. As a result, books were rare and expensive.

The man who first invented a practical method of printing was a German named Johannes Gutenberg. In about 1438, he discovered a way of making metal type – small metal blocks with a single letter cut into one end – from which books could be printed. The idea behind Gutenberg's invention was not new. The Chinese had also invented a method of printing using movable type, but the Chinese language has thousands of characters and the system was impractical.

Gutenberg also invented an ink which would cling to the type and adapted a wine press to press paper evenly onto the inked type. By about 1450 he had perfected his techniques and began printing the first books. The first large book he printed was a Bible. Other printers, such as William Caxton in England, soon set up printing works. As a result, books became cheaper and much more common.

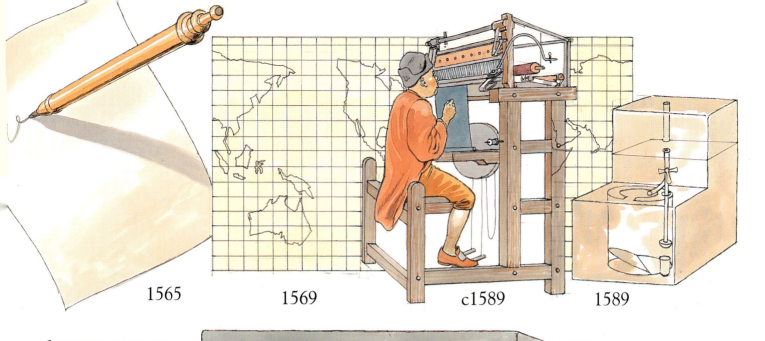

1565

1569

c1589

1589

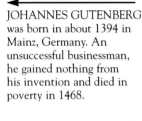

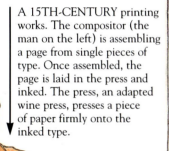

JOHANNES GUTENBERG was born in about 1394 in Mainz, Germany. An unsuccessful businessman, he gained nothing from his invention and died in poverty in 1468.

A 15TH-CENTURY printing works. The compositor (the man on the left) is assembling a page from single pieces of type. Once assembled, the page is laid in the press and inked. The press, an adapted wine press, presses a piece of paper firmly onto the inked type.

1400 The first mechanical clocks were made around 1290 in English and Italian monasteries. Alarm clocks appeared between 1350 and 1400. They were used to wake sleeping monks.

1452 Birth of Leonardo da Vinci, painter, sculptor, scientist and inventor. His designs for helicopters and submarines were centuries ahead of their time.

1498 The toothbrush, with bristles at right angles to the handle, was first described in a Chinese encylopedia.

1538 The first recorded use of a diving bell was in 1538 in Toledo, Spain. The bell shown was designed by English scientist Edmund Halley in 1717.

1565 The pencil was invented by Swiss doctor Konrad von Gesner in Switzerland. It had a centre of pure graphite (a soft black form of carbon) covered with wood.

1569 Flemish geographer Gerard Mercator produced a map which treated the globe as a cylinder rolled out flat. Navigators were able to plot their course as straight lines on the new map.

c1589 English clergyman William Lee invented the first mechanical knitting machine around 1589. The machine could produce material as fine as silk ten times faster than handknitters.

1589 The flush toilet was invented by John Harington, the godson of Queen Elizabeth I of England. In 1589 the new toilet was installed in Harington's home in Somerset, in the west of England. It did not become popular until Victorian times.

c1590

c1593

c1610

SEEING NEAR AND FAR

A simple pair of spectacles made in Italy around 1500. They were clamped to the nose and used for reading.

HOW a telescope works. In a refracting telescope (top) the light rays pass through two lenses before reaching the eye. In a reflecting telescope (bottom) light is reflected from two mirrors to a second before it reaches the eyepiece.

The first spectacle-maker was Salvino Armati of Florence in Italy. He made what he called 'little discs for the eyes' in 1286. The discs were convex lenses – circles of glass with surfaces which bulged outwards – which helped long-sighted people see small objects and writing. Lenses were the key to two very important inventions: the microscope and the telescope.

A single convex lens is a simple microscope; it can magnify small objects so that they look larger. About 1590 Dutchman Antonie von Leeuwenhoek used a small, powerful lens to reveal a whole new world of creatures never before seen. The first microscope using more than one lens was made by Dutch spectacle-maker Hans Jansen and his son Zacharias in 1590. It was called a compound microscope.

In 1608 a Dutchman called Hans Lippershey happened to look at a distant rooftop through two lenses. He was surprised to see that the rooftop seemed much closer than it really was. Quickly he mounted the lenses at the ends of a hollow tube, so making the first telescope. In 1609, Italian scientist Galileo Galilei heard stories of the new instrument and made a telescope of his own. He used it to look at the stars and planets and made many discoveries.

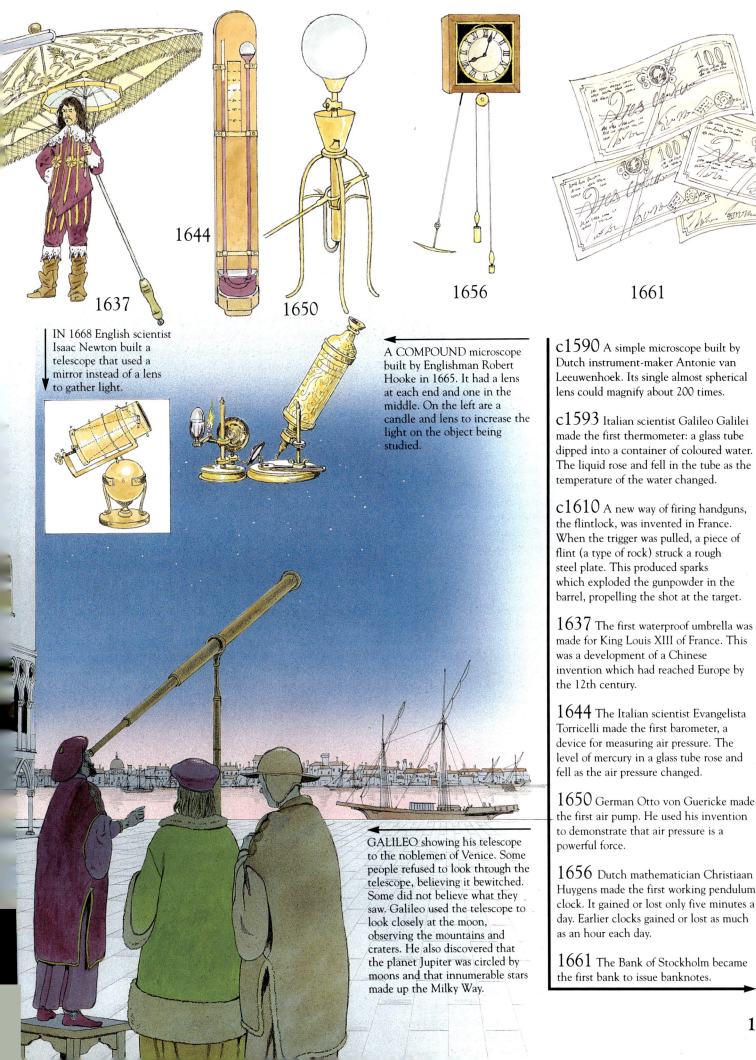

1637

1644

1650

1656

1661

IN 1668 English scientist Isaac Newton built a telescope that used a mirror instead of a lens to gather light.

A COMPOUND microscope built by Englishman Robert Hooke in 1665. It had a lens at each end and one in the middle. On the left are a candle and lens to increase the light on the object being studied.

GALILEO showing his telescope to the noblemen of Venice. Some people refused to look through the telescope, believing it bewitched. Some did not believe what they saw. Galileo used the telescope to look closely at the moon, observing the mountains and craters. He also discovered that the planet Jupiter was circled by moons and that innumerable stars made up the Milky Way.

c1590 A simple microscope built by Dutch instrument-maker Antonie van Leeuwenhoek. Its single almost spherical lens could magnify about 200 times.

c1593 Italian scientist Galileo Galilei made the first thermometer: a glass tube dipped into a container of coloured water. The liquid rose and fell in the tube as the temperature of the water changed.

c1610 A new way of firing handguns, the flintlock, was invented in France. When the trigger was pulled, a piece of flint (a type of rock) struck a rough steel plate. This produced sparks which exploded the gunpowder in the barrel, propelling the shot at the target.

1637 The first waterproof umbrella was made for King Louis XIII of France. This was a development of a Chinese invention which had reached Europe by the 12th century.

1644 The Italian scientist Evangelista Torricelli made the first barometer, a device for measuring air pressure. The level of mercury in a glass tube rose and fell as the air pressure changed.

1650 German Otto von Guericke made the first air pump. He used his invention to demonstrate that air pressure is a powerful force.

1656 Dutch mathematician Christiaan Huygens made the first working pendulum clock. It gained or lost only five minutes a day. Earlier clocks gained or lost as much as an hour each day.

1661 The Bank of Stockholm became the first bank to issue banknotes.

c1670 1675 1679 1701

STEAM POWER

The first steam engine, the aeolipile, was built in Egypt around 150 BC. Steam flowed from jets on a copper globe and drove the globe around.

Steam power was first demonstrated in Egypt over 2,100 years ago. An inventor named Ctesibius made a machine called an aeolipile which was spun round by steam power. However, the first useful steam engine was built in 1712 by Thomas Newcomen, an English blacksmith. Newcomen's engine was used to pump water from flooded mines. It had a piston which moved up and down inside a tube or cylinder. The piston rose as steam flowed into the cylinder. When the piston was at the top of the cylinder, water was sprayed into the cylinder, condensing the steam. This created a vacuum under the piston and air pressure forced the piston down.

In 1763 a Scottish instrument-maker, James Watt, reasoned that the engine would produce more power if the steam was condensed in a separate chamber connected to the cylinder. This would mean that the cylinder did not have to be reheated after every piston movement. Watt began a series of experiments to test his ideas. It took over a year, but eventually Watt built an improved engine and soon it was installed in factories and mines across Britain.

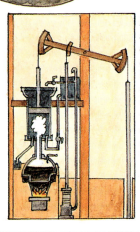

Newcomen's steam engine was called a beam engine. A large beam rocked up and down as the piston moved in the cylinder.

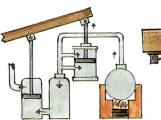

WATT'S steam engine used a separate condenser for cooling the steam. This allowed the cylinder to stay hot, so less heat was wasted.

JAMES WATT in his workshop. His engine was the first true steam engine, as it used the pressure of steam to drive the piston along the cylinder. Over the years, Watt gradually improved his engine. In 1782 he made the engine 'double acting' – steam pushed the piston down the cylinder as well as up. He also invented a 'governor' which controlled the speed of the engine. Watt introduced the term 'horsepower', comparing the power of an average horse with the power of his engine. The unit used to measure electrical power is called the watt after him.

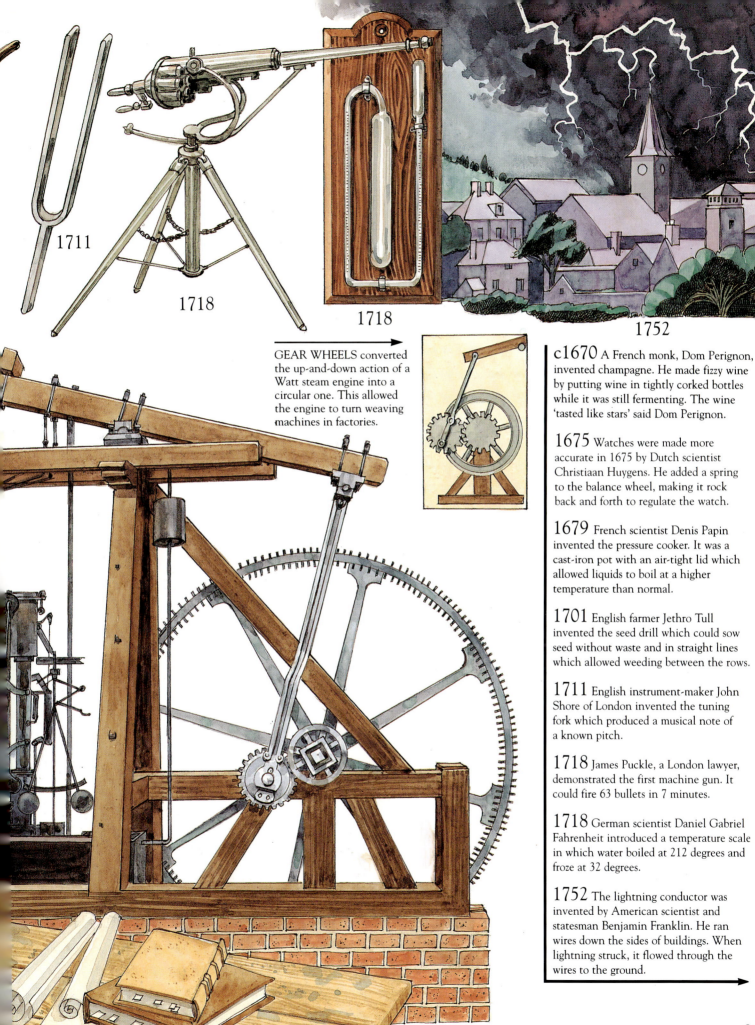

1711

1718

1718

1752

GEAR WHEELS converted the up-and-down action of a Watt steam engine into a circular one. This allowed the engine to turn weaving machines in factories.

c1670 A French monk, Dom Perignon, invented champagne. He made fizzy wine by putting wine in tightly corked bottles while it was still fermenting. The wine 'tasted like stars' said Dom Perignon.

1675 Watches were made more accurate in 1675 by Dutch scientist Christiaan Huygens. He added a spring to the balance wheel, making it rock back and forth to regulate the watch.

1679 French scientist Denis Papin invented the pressure cooker. It was a cast-iron pot with an air-tight lid which allowed liquids to boil at a higher temperature than normal.

1701 English farmer Jethro Tull invented the seed drill which could sow seed without waste and in straight lines which allowed weeding between the rows.

1711 English instrument-maker John Shore of London invented the tuning fork which produced a musical note of a known pitch.

1718 James Puckle, a London lawyer, demonstrated the first machine gun. It could fire 63 bullets in 7 minutes.

1718 German scientist Daniel Gabriel Fahrenheit introduced a temperature scale in which water boiled at 212 degrees and froze at 32 degrees.

1752 The lightning conductor was invented by American scientist and statesman Benjamin Franklin. He ran wires down the sides of buildings. When lightning struck, it flowed through the wires to the ground.

1761

1757

1764

1770

SANDWICHES AND STEAMERS

In 1790, the first steamboat service was operated by American John Fitch on the Delaware River in the eastern USA.

The first steamboat was built in France in 1783 by the Marquis Joffroy d'Abans. Like most early steamboats, its engines turned a paddle-wheel which drove the boat along. However, the paddle-wheel was not efficient and was eventually replaced by the propeller.

The propeller was invented by an English farmer, Francis Pettit Smith, in 1835. To prove that the propeller was better than the paddle-wheel, Pettit Smith made a small steamship with a wooden propeller and sailed it around the coast of Kent in southern England.

At first, steam ships had to carry large amounts of fresh water for the boiler and lots of coal as fuel. This meant that there was little room for cargo. Then, in 1834, English engineer Samuel Hall invented a condenser which was able to turn the waste steam from the engine back into water. This meant the water could be reused and less was needed. Around the same time, English engineer Isambard Kingdom Brunel realized that the solution to the fuel problem was to build very large ships. These giant ships had room for a large cargo as well as plenty of fuel. Brunel built several large steamers to prove his point.

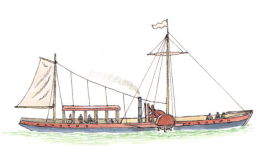

THE *Clermont*, the first passenger-carrying paddle-steamer. It was built by American Robert Fulton in 1807. The *Clermont* steamed up the Hudson River between New York and Albany.

IN 1838, two steamships raced across the Atlantic. The idea was to prove that large steamships were the best vessels to carry cargo on long sea voyages. Taking part was the giant *Great Western*, specially built to cross the Atlantic by Isambard Kingdom Brunel, and the much smaller *Sirius*. On the voyage, the *Sirius* ran out of fuel. Cabin doors, furniture and deck planks were thrown into the boiler to keep her going. She took 18 days to make the journey. The *Great Western* made the crossing in 15 days and had plenty of coal left at the end of the journey.

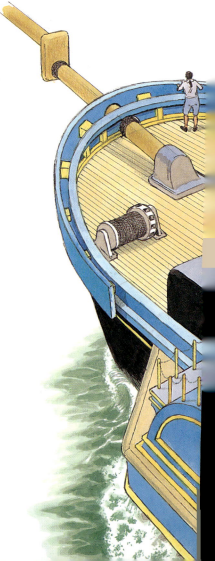

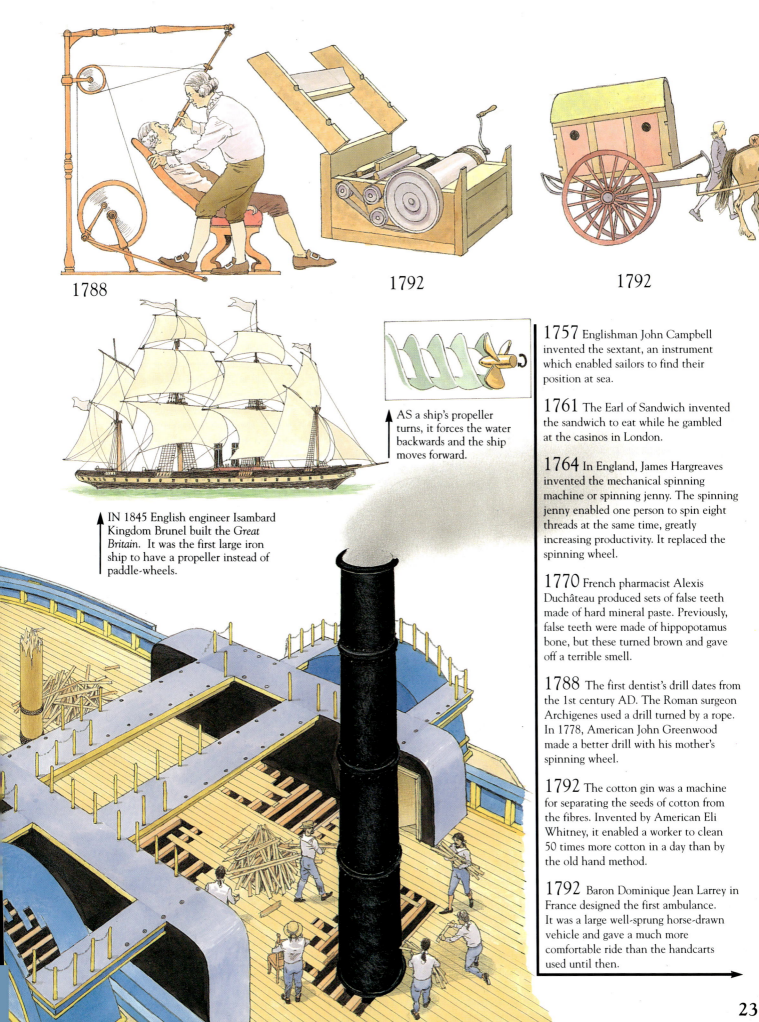

1788

1792

1792

AS a ship's propeller turns, it forces the water backwards and the ship moves forward.

IN 1845 English engineer Isambard Kingdom Brunel built the *Great Britain*. It was the first large iron ship to have a propeller instead of paddle-wheels.

1757 Englishman John Campbell invented the sextant, an instrument which enabled sailors to find their position at sea.

1761 The Earl of Sandwich invented the sandwich to eat while he gambled at the casinos in London.

1764 In England, James Hargreaves invented the mechanical spinning machine or spinning jenny. The spinning jenny enabled one person to spin eight threads at the same time, greatly increasing productivity. It replaced the spinning wheel.

1770 French pharmacist Alexis Duchâteau produced sets of false teeth made of hard mineral paste. Previously, false teeth were made of hippopotamus bone, but these turned brown and gave off a terrible smell.

1788 The first dentist's drill dates from the 1st century AD. The Roman surgeon Archigenes used a drill turned by a rope. In 1778, American John Greenwood made a better drill with his mother's spinning wheel.

1792 The cotton gin was a machine for separating the seeds of cotton from the fibres. Invented by American Eli Whitney, it enabled a worker to clean 50 times more cotton in a day than by the old hand method.

1792 Baron Dominique Jean Larrey in France designed the first ambulance. It was a large well-sprung horse-drawn vehicle and gave a much more comfortable ride than the handcarts used until then.

1815
1816
1818
1823

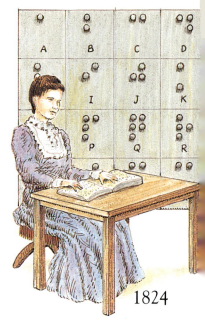

1824

MAKING ELECTRICITY

A machine made in 1713 to produce static electricity. Turning the handle produced an electric charge on the glass globe at the top.

When you run a plastic comb through your hair, you will sometimes see small sparks jump from the comb. These sparks are due to static electricity. Static electricity can also be produced by rubbing a cloth against glass; early electricity generators worked in this way.

A better way of producing electricity was discovered in 1800 when Italian scientist Alessandro Volta invented the battery. The battery produced a steady electric current. One scientist who experimented with the electric battery was Englishman Michael Faraday. He made several important inventions in 1831, including the electric generator and the transformer.

Faraday's generator produced electric current by moving a copper disc near a magnet. It only produced a small current. In 1832 Frenchman Hyppolyte Pixii made the first practical generator and the modern electrical industry was born. The first power station, driven by water, generated electricity to light the streets of Godalming in south-east England in 1881.

Faraday's other invention, the transformer, was also adopted by the electrical industry. It was used to increase the voltage of an electric current. High voltages were easier to transmit long distances over wires.

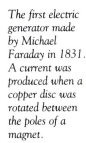

ALESSANDRO VOLTA and his battery. It was made of discs of copper, zinc and cloth soaked in brine.

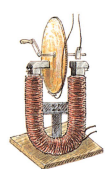

The first electric generator made by Michael Faraday in 1831. A current was produced when a copper disc was rotated between the poles of a magnet.

MICHAEL FARADAY with the apparatus he used to show how electricity flowing in one coil can produce electricity in another coil. This discovery led to the electrical transformer.

A GENERATOR built by American inventor Thomas Alva Edison in 1880. In 1882 Edison built an electricity station in Pearl Street, New York, which had eight generators. The station could only provide electricity to nearby homes because Edison used direct current. Direct current, in which electricity flows in one direction along the wire, cannot be transmitted long distances. Alternating current, in which electricity reverses its direction of flow many times a second, was developed by Croatian-born US inventor Nikola Tesla in the 1800s. Alternating current can be transmitted long distances with the help of the transformer.

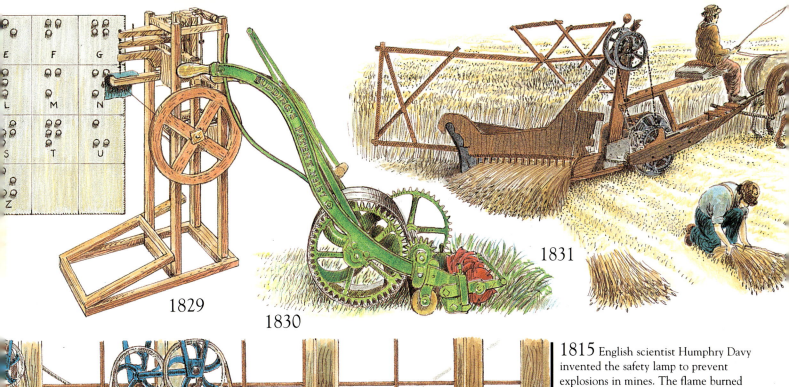

1829

1830

1831

1815 English scientist Humphry Davy invented the safety lamp to prevent explosions in mines. The flame burned inside a metal gauze shield and could not ignite explosive gases in the mine.

1816 Johann Maelzel invented the metronome, a device which indicates the exact tempo musicians should follow when playing a piece.

1818 Developing the pioneering work of the Frenchman Nicolas Appert in preserving food in glass and tin containers, an English company began supplying canned food to the Royal Navy. In 1938 a tin of veal and peas canned in 1818 was opened. It was still fresh.

1823 Scotsman Charles MacIntosh produced a waterproof cloth that he used to make raincoats. They became known as 'Macintoshes'.

1824 Blind since the age of three, Frenchman Louis Braille invented an alphabet for the blind using raised dots.

1829 French tailor Barthélemy made the first sewing machine. He set up a factory with 80 machines making army uniforms.

1830 English textile worker Edwin Budding invented the first successful lawn mower. The mower was based on a cloth-cutting machine and today's cylinder mowers look very like it.

1831 In 1826 Scottish minister Patrick Bell had invented a reaper to harvest grain crops. It was pulled by two horses. In 1831, American Cyrus McCormick invented a more efficient machine.

1836

1840

1846

CAMERAS TO CAN OPENERS

The camera obscura, a forerunner of the camera. The image formed by the lens at the front was outlined by hand.

A FRENCH camera, 1864. The image was formed on a glass plate. The photographer carried chemicals to process the plate.

WILLIAM FOX TALBOT made photographs on film, a special transparent paper.

A PHOTOGRAPHER'S studio in 1888. By this time, only a fraction of a second was needed to take a photograph. The camera, however, was still large and needed a sturdy tripod stand to keep it steady. Reflectors directed light onto the subject. When working outdoors, a photographer needed an assistant to help carry all the equipment.

The simplest type of camera, the pinhole camera, was invented about 900 years ago. It was a small box with a hole in one end to let in light. The light made an image on the back of the box. The first photograph was taken in 1816 by a Frenchman called Nicephore Niepce using a pinhole camera. At the back of the camera was a metal plate coated with a thin layer of bitumen and oil. After about eight hours, the bitumen became hard where the light was strongest. Niepce washed off the soft bitumen and a picture was formed by the bitumen that remained.

In 1835, another Frenchman, Louis Daguerre, invented a better process. It was quicker and produced a sharper image. Daguerre used a metal plate coated with silver, which darkened when exposed to light. It took half an hour to make a single daguerreotype, as the images were known.

Also in 1835 an English photographer, William Fox Talbot, discovered how to make photographs on special paper that darkened rapidly when light shone on it. This meant that photographs could be taken much more quickly – in about two minutes. Fox Talbot's discoveries began modern photography.

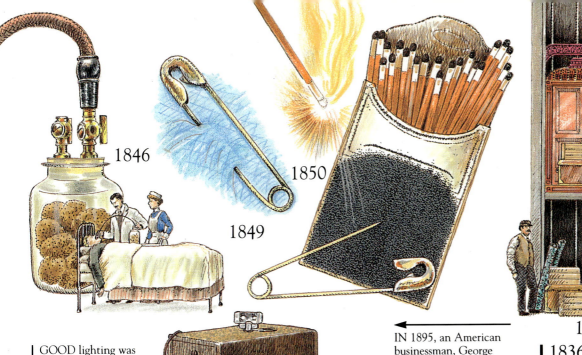

1846

1849

1850

1854   1855

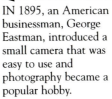

GOOD lighting was essential and studios had large windows. If extra light was required, magnesium was burned to produce a bright white light.

IN 1895, an American businessman, George Eastman, introduced a small camera that was easy to use and photography became a popular hobby.

1836 Americans Elisha Collier and Artemis Wheeler had invented a revolver which could fire six shots in 1818. In 1836 American gunsmith Samuel Colt simplified the gun and mass-produced it.

1840 In May 1840, the British post office introduced the first postage stamp, the Penny Black. It had a picture of Queen Victoria's head.

1846 English rubber manufacturer Thomas Hancock made a set of carriage tyres of solid rubber fixed to an iron rim around the wooden wheel.

1846 American dentist William Morton used ether as an anaesthetic (pain killer). The ether was breathed from a glass jar filled with ether-soaked sponges.

1849 American Walter Hunt invented the safety pin. He sold the rights to his invention to pay a $15 debt.

1850 The Swedish inventer Johan Lundstrom made the safety match. The matches caught fire only when struck on a special surface on the match box.

1854 American inventor Elisha Otis invented the elevator or lift. It had a device which stopped it falling if the supporting cable snapped. To demonstrate how safe his invention was, Otis stood in the elevator cage while the cable was cut.

1855 Robert Yeates, an Englishman, invented the first simple can opener. It became popular ten years later when it was given away with tins of canned beef. Before Yeates' invention, a hammer and chisel were needed to open cans.

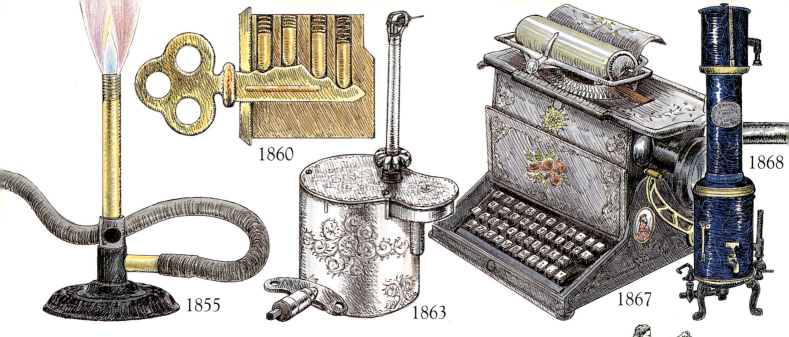

1855

1860

1863

1867

1868

Bicycles to Light Bulbs

The draisienne or 'running machine' of 1817. On good roads, it was four times faster than a horse-drawn coach.

The first practical bicycle was made in 1817 by a German, Karl von Drais. His invention, which he called a draisienne, had a wooden body set above two wheels in line. The rider moved along by pushing on the ground, first with one foot and then with the other.

Around 1861, a Frenchman, Pierre Michaux, and his son Ernest, fitted pedals to the front wheel of a draisienne so that the rider did not have to 'paddle' along using the feet. Michaux called his machines velocipedes or boneshakers.

In 1870 James Starley of Coventry, England, introduced the 'penny farthing' bicycle. This had a very large front wheel, and a small back wheel. The penny farthing was fast but difficult to ride. The rider could easily be thrown over the handlebars onto the ground. This disadvantage led to the development of the safety bicycle.

The first safety bicycle was made by Harry Lawson and ridden around the streets of Brighton, England, in 1873. The two wheels on a safety bicycle were the same size. The pedals were connected to the back wheel by a chain. The brake, too, was attached to the back wheel. These features meant that the safety bicycle was easy and safe to ride.

MOUNTING a penny farthing was difficult. The rider had to step quickly into the saddle and start pedalling furiously before the bike fell over.

PERCHED high above the ground, the penny-farthing rider could reach speeds of 32 km/h. People complained bitterly about the new machines, saying speeding riders frightened horses, excited dogs and scared children.

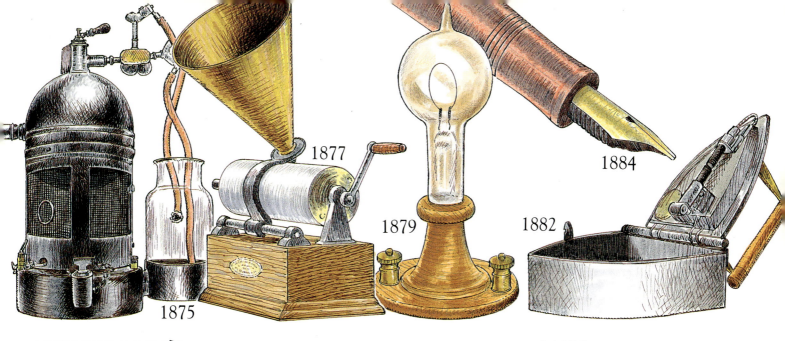

1877

1875

1879

1882

1884

IN 1839, a Scottish blacksmith, called Kirkpatrick Macmillan, built the first bicycle with pedals. The pedals were attached to the back wheel by rods. In 1842 Macmillan made a 224-kilometre journey on it. Unfortunately, Macmillan's ideas were forgotten and his bicycle was not developed further.

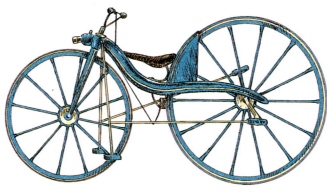

↑ THE velocipede or boneshaker bicycle of 1861. To everyone's surprise, it was possible to keep the machine upright even when both feet were off the ground.

↑ THE Rover safety bicycle became extremely popular. It was easier to handle and more comfortable than the penny farthing and there was less danger of falling off.

1855 German scientist Robert Bunsen introduced the Bunsen burner which produced a very hot flame.

1860 Linus Yale Jr of Philadelphia, an American locksmith, invented the pin-tumbler lock, which remains in use today.

1863 Englishman George Fellows Harrington invented a hand-held clockwork dental drill.

1867 The first practical typewriter was built in the USA by Christopher Scholes and Carlos Glidden.

1868 The gas water heater or geyser was invented by London decorator Waddy Maughan in 1868.

1875 An improved version of the carbolic acid spray which Scottish surgeon James Lister had invented in 1867 to kill germs during surgical operations.

1877 American inventor Thomas Alva Edison invented the phonograph, or 'talking machine'. The first words recorded were 'Mary had a little lamb'.

1879 Thomas Alva Edison made the first successful light bulb. It had a carbon filament in a vacuum; air was removed from the bulb to prevent the filament from burning.

1882 Henry Seely of New York made the first electric iron.

1884 American insurance salesman Lewis Waterman became annoyed by dripping pens. He solved the problem by producing the first successful fountain pen.

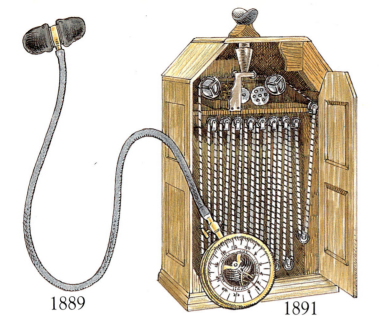

1889

1891

1892

1892

SAMUEL MORSE sent the first 'Morse code' message from Washington to Baltimore in 1844. His receiver printed out the message in dots and dashes as it was being received.

Semaphore to Telephones

The semaphore telegraph invented by Frenchman Claude Chappé in the 1790s. Messages were sent by changing the position of the arms.

Fast long-distance communication began during the French Revolution, in the 1790s. In 1794 a French merchant called Claude Chappé used a semaphore telegraph to send messages between the French army at Lille and Paris. A message could be sent to Paris 240 kilometres away in two minutes. Semaphore telegraphs were very popular in the early 1800s, until the electric telegraph was invented.

The first electric telegraph, which sent a message along a wire, was built in England in 1837 by Charles Wheatstone and William Cooke. An American inventor, Samuel Morse, improved the telegraph in 1840. His system used a buzzer that made a sound; messages were sent as a code of short and long buzzes. This code was later called Morse code.

The telephone was invented by Scottish-born US inventor Alexander Graham Bell. On 10 March, 1876, he sent his first telephone message to his assistant Mr Watson in another room. The message was 'Mr Watson, come here; I want to see you'. In 1877 Thomas Alva Edison made an improved telephone that could be heard over longer distances.

It was difficult to hear what was said on an early Bell telephone. Some models had two earpieces.

30

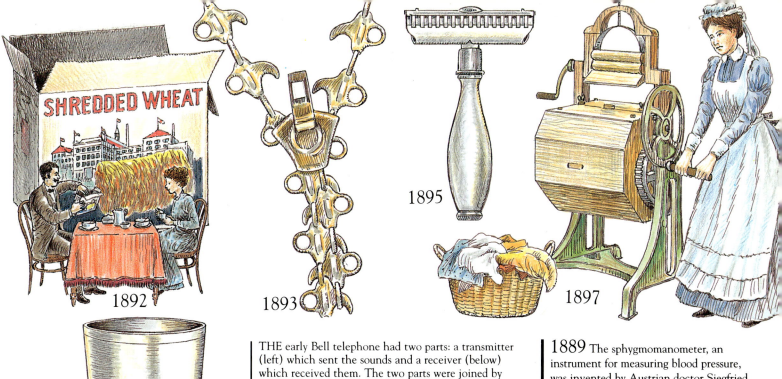

SHREDDED WHEAT

1892

1893

1895

1897

THE early Bell telephone had two parts: a transmitter (left) which sent the sounds and a receiver (below) which received them. The two parts were joined by wires. The transmitter and receiver each had a thin piece of metal, called a reed, near an electric magnet. When someone spoke near the transmitter, the reed vibrated back and forth. This movement made the magnet vibrate, causing an electric current in the wire to the receiver. The electric current made the reed in the receiver move, reproducing the words spoken into the transmitter.

1889 The sphygmomanometer, an instrument for measuring blood pressure, was invented by Austrian doctor Siegfried von Basch in 1881. The instrument shown is a type of sphygmomanometer developed from it.

1891 US inventor Thomas Alva Edison invented the kinetoscope, the first motion picture system. It contained a strip of film which moved past the eyepiece at high speed, creating the moving pictures.

1892 Coca Cola was invented by doctor John Pemberton of Atlanta, USA. The drink was first sold as 'an esteemed brain tonic' at Jacob's Pharmacy in Atlanta for 5 cents a glass.

1892 Scottish scientist James Dewar invented the vacuum flask to keep liquids hot or cold.

1892 Henry Perky, a lawyer from Denver, Colorado, USA, invented shredded wheat, the first modern breakfast cereal. The world's most popular breakfast cereal, corn flakes, was invented in 1902 by William Kellogg.

1893 The zip was invented by Whitcomb Judson, an engineer from Chicago, USA. However, the first zips were unreliable and often burst open.

1895 Travelling salesman King Camp Gillette, of Fond du Lac, Wisconsin, USA, invented the safety razor. The new razor had disposable double-edged blades.

1897 A washing machine with a mangle attached for squeezing the washing dry. The first washing machine was introduced in 1884.

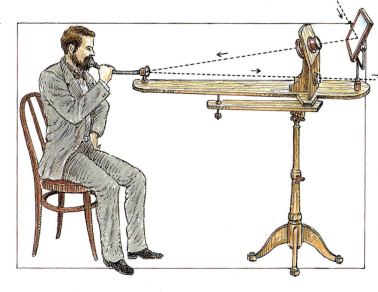

IN 1880 Alexander Graham Bell invented the photophone, which transmitted sounds using light beams. Sounds spoken into the transmitter moved a mirror. This altered a beam of light shining on the mirror. The receiver was able to detect the changes in the light beam and reproduce the sounds spoken.

1895

1897

1900

1901

1904

Petrol Engines Appear

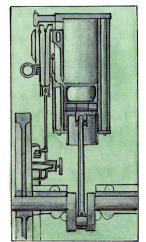

An 1890s steam engine modified to run on petrol. These engines could not produce the power of a four-stroke engine.

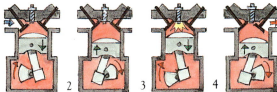

1   2   3   4

THE four-stroke engine cycle. 1 Induction stroke. The piston moves down, sucking fuel and air into the cylinder. 2 Compression stroke. The piston moves up, squeezing the fuel and air mixture. 3 Power stroke. A spark ignites the mixture. The exploding fuel forces the piston down. 4 Exhaust stroke. The piston moves up, forcing the burnt fuel out of the cylinder.

Although the petrol engine is sometimes called the 'internal combustion engine', any engine that burns fuel inside it ('combustion' means burning) is an internal combustion engine.

The first working internal combustion engine was built in 1859 by Belgian inventor Etienne Lenoir. It used an explosive gas as fuel. A German engineer, Nikolaus Otto, invented a better gas engine in 1876. It was called the four-stroke engine because it used four movements of the piston to produce its power.

The next major advance was made by German engineers Gottlieb Daimler and Karl Benz. They modified the four-stroke engine to run on petrol. This was a revolutionary invention because it meant the engine was portable. Gas engines had to be stationary, fixed to the gas supply.

In 1878 Dugald Clerk, a Scottish engineer, produced the first two-stroke engine. The two-stroke engine was simple, lighter and easier to make; but it was noisy and not as powerful as the four-stroke engine.

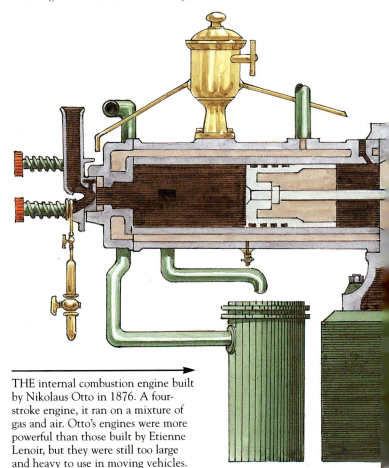

THE internal combustion engine built by Nikolaus Otto in 1876. A four-stroke engine, it ran on a mixture of gas and air. Otto's engines were more powerful than those built by Etienne Lenoir, but they were still too large and heavy to use in moving vehicles.

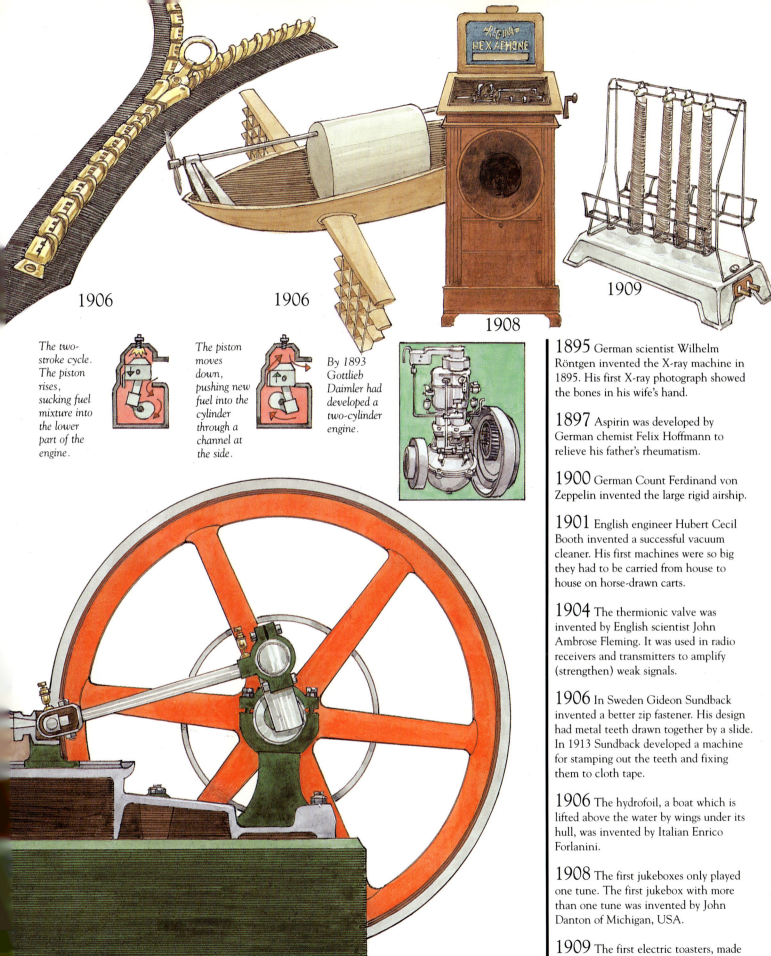

1906

1906

1908

1909

The two-stroke cycle. The piston rises, sucking fuel mixture into the lower part of the engine.

The piston moves down, pushing new fuel into the cylinder through a channel at the side.

By 1893 Gottlieb Daimler had developed a two-cylinder engine.

1895 German scientist Wilhelm Röntgen invented the X-ray machine in 1895. His first X-ray photograph showed the bones in his wife's hand.

1897 Aspirin was developed by German chemist Felix Hoffmann to relieve his father's rheumatism.

1900 German Count Ferdinand von Zeppelin invented the large rigid airship.

1901 English engineer Hubert Cecil Booth invented a successful vacuum cleaner. His first machines were so big they had to be carried from house to house on horse-drawn carts.

1904 The thermionic valve was invented by English scientist John Ambrose Fleming. It was used in radio receivers and transmitters to amplify (strengthen) weak signals.

1906 In Sweden Gideon Sundback invented a better zip fastener. His design had metal teeth drawn together by a slide. In 1913 Sundback developed a machine for stamping out the teeth and fixing them to cloth tape.

1906 The hydrofoil, a boat which is lifted above the water by wings under its hull, was invented by Italian Enrico Forlanini.

1908 The first jukeboxes only played one tune. The first jukebox with more than one tune was invented by John Danton of Michigan, USA.

1909 The first electric toasters, made by the General Electric Company, New York, USA, went on sale.

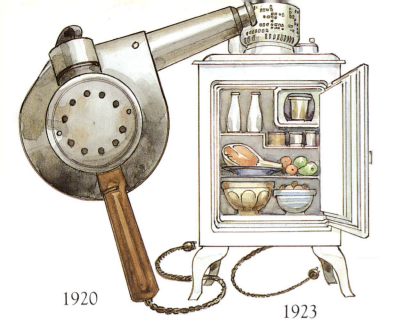

1920

1923

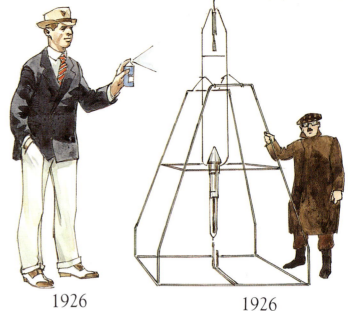

1926

1926

POWERED VEHICLES

Nicolas Cugnot's 1769 steam wagon. Unfortunately it was difficult to steer and crashed on its first journey.

The first powered vehicles on the roads were steam-powered. In 1769, a Frenchman, Nicolas Cugnot, built a three-wheeled 'steam wagon' to pull heavy guns. The earliest motorcycles were draisiennes driven by small steam engines.

The first steam locomotive was built by Cornish engineer Richard Trevithick in 1803 to carry coal for the Coalbrookdale Ironworks in Shropshire, England. Another railway pioneer was George Stephenson. With his son Robert, George Stephenson started the world's first public railway in 1825, between the towns of Stockton and Darlington in northern England.

In 1885 German engineer Gottlieb Daimler fitted a petrol engine to a wooden bicycle, so making the first petrol-driven motorcycle. The following year he fitted a petrol engine to a horse carriage to make the first automobile. It was called a 'horseless carriage'. At the same time, Karl Benz, another German engineer, was also making petrol-engined cars. Benz's car was lighter than Daimler's, and only had three wheels. Benz was the first person to make cars for sale, starting a car factory in 1894.

GOTTLIEB DAIMLER riding in his 'horseless carriage'. It made its first public run on 3 July, 1886, at Mannheim, Germany, travelling just over 0.8 kilometre at a speed of 15 km/h.

THE FIRST trials of the Benz car were not encouraging. The car either refused to start, or started so suddenly that the passengers were thrown out. When it got going, the car could cruise at a speed of 13 km/h. The local newspaper described the car as 'useless, ridiculous and indecent'.

1927

1928

1928

1931

GEORGE STEPHENSON'S locomotive, *Rocket*, won a competition held in 1829 to find the best steam engine. It reached a speed of 46 km/h.

THE FIRST petrol-driven motorcycle. Made by Daimler in 1885, its critics described it as 'diabolical and dangerous'.

1920 The first hand-held hairdryer was made by the Racine Universal Motor Company in the USA. A small electric motor blew air over a heated filament.

1923 The electric refrigerator was pioneered by Swedish engineers Balzer von Platen and Carl Munters with their 'Electrolux' model.

1926 Norwegian Erik Rotheim invented the aerosol can which produced a fine spray of liquid or powder.

1926 Robert Goddard launched the first liquid-fuelled rocket near Auburn, Massachusetts, USA. It reached a height of 12 metres.

1927 The first films had no soundtrack. In 1927 *The Jazz Singer*, starring Al Jolson, was released. It became the first commercially successful 'talkie'. Jolson's first words in the film were 'Wait a minute, wait a minute. You ain't heard nothin' yet!'

1928 Alexander Fleming, working at St Mary's Hospital, London, discovered penicillin, the first antibiotic (germ-killing substance derived from a living source).

1928 American professor Philip Drinker invented the iron lung to help patients with diseased lungs to breathe. It was tested on a young girl at Boston Hospital on 12 October, 1928.

1931 Retired US army officer Jacob Schick sold the first electric razor for $31. A small electric motor drove a row of blades backwards and forwards behind a slotted guard.

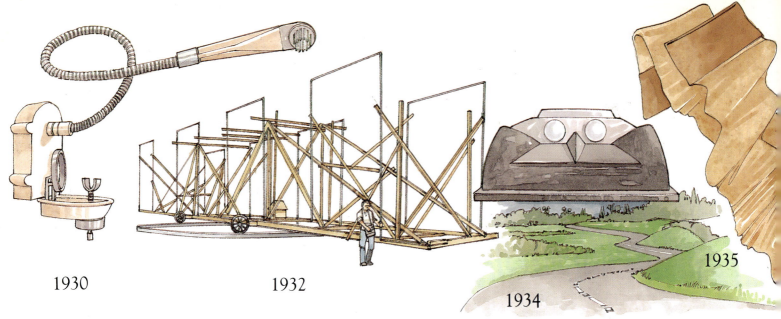

1930

1932

1934

1935

FLYING MACHINES

The Montgolfier hot-air balloon. On its first flight, in September 1783, it had a duck, a sheep and a cockerel aboard.

The first successful flying machines were hot-air balloons. On 21 November, 1783, Pilâtre de Rozier and the Marquis d'Arlandes took a 25-minute flight over Paris, France, in a hot-air balloon made by the Montgolfier brothers, Joseph and Etienne. The world's first airship was flown in 1852 by Frenchman Henri Giffard. The airship was a cigar-shaped balloon with a propeller turned by a steam engine. It was the first steerable flying machine.

The next important figure in flying history was Otto Lilienthal, a German engineer. Between 1891 and 1896, Lilienthal made hundreds of flights using a hang-glider that he designed. He discovered the best wing shape and how to control the glider in the air.

Lilienthal's work was carried on by the Wright brothers, Wilbur and Orville, in the USA. By 1903 they had made over 1000 flights in gliders and were ready for the next step: powered flight. They built a small powerful petrol engine and added it to their glider. On 17 December, 1903, the airplane, *Flyer 1*, made the first powered flight at Kitty Hawk, North Carolina.

THE FIRST public parachute jump from a balloon was made over Paris by André-Jacques Garnerin in October 1797.

Otto Lilienthal controlled his glider by leaning from side to side, as a modern hang-glider pilot does. He made flights of 225 metres.

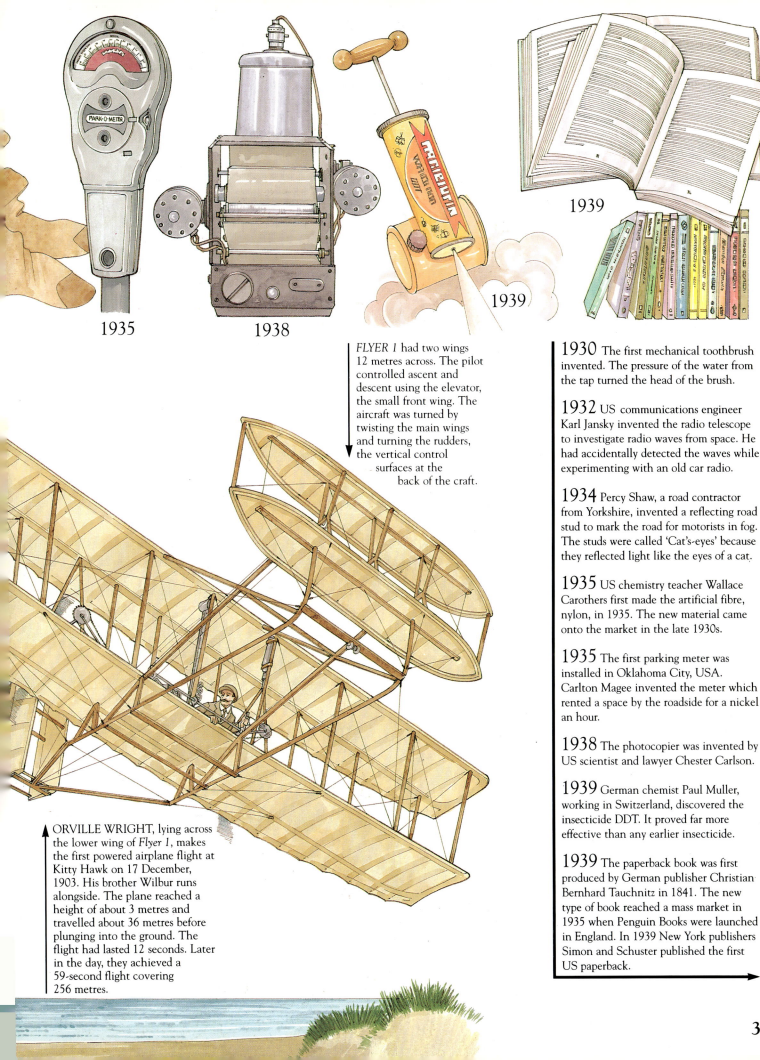

1935

1938

1939

1939

FLYER 1 had two wings 12 metres across. The pilot controlled ascent and descent using the elevator, the small front wing. The aircraft was turned by twisting the main wings and turning the rudders, the vertical control surfaces at the back of the craft.

ORVILLE WRIGHT, lying across the lower wing of *Flyer 1*, makes the first powered airplane flight at Kitty Hawk on 17 December, 1903. His brother Wilbur runs alongside. The plane reached a height of about 3 metres and travelled about 36 metres before plunging into the ground. The flight had lasted 12 seconds. Later in the day, they achieved a 59-second flight covering 256 metres.

1930 The first mechanical toothbrush invented. The pressure of the water from the tap turned the head of the brush.

1932 US communications engineer Karl Jansky invented the radio telescope to investigate radio waves from space. He had accidentally detected the waves while experimenting with an old car radio.

1934 Percy Shaw, a road contractor from Yorkshire, invented a reflecting road stud to mark the road for motorists in fog. The studs were called 'Cat's-eyes' because they reflected light like the eyes of a cat.

1935 US chemistry teacher Wallace Carothers first made the artificial fibre, nylon, in 1935. The new material came onto the market in the late 1930s.

1935 The first parking meter was installed in Oklahoma City, USA. Carlton Magee invented the meter which rented a space by the roadside for a nickel an hour.

1938 The photocopier was invented by US scientist and lawyer Chester Carlson.

1939 German chemist Paul Muller, working in Switzerland, discovered the insecticide DDT. It proved far more effective than any earlier insecticide.

1939 The paperback book was first produced by German publisher Christian Bernhard Tauchnitz in 1841. The new type of book reached a mass market in 1935 when Penguin Books were launched in England. In 1939 New York publishers Simon and Schuster published the first US paperback.

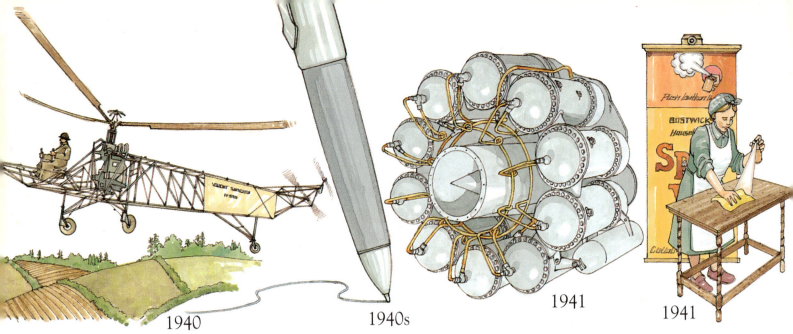

1940

1940s

1941

1941

THE EARLY 20TH CENTURY

The radio receiver used by Heinrich Hertz was a simple loop of wire. A spark appeared in the gap when a radio wave was received.

Radio and television do not use wires to carry messages. They use a type of wave called an electromagnetic wave. Heinrich Hertz, a German physicist, first detected these waves in 1887.

Italian Guglielmo Marconi showed that radio waves could carry messages to the most distant parts of the world. In 1901, Marconi sent a radio signal from Cornwall, England, to Newfoundland in Canada, a distance of 3,520 kilometres. The message was the single letter S in Morse code. Twenty-three years later, in 1924, Marconi succeeded in sending a radio message from England to Australia.

The first person to transmit television pictures was Scottish engineer John Logie Baird. In October 1925, Baird succeeded in producing the first television picture of a person, a 15-year-old boy called Billy Taynton who was paid 12 pence for his help. In 1929 Baird persuaded the British Broadcasting Corporation (BBC) to transmit the world's first television service in London. At the time there were only 100 television sets to receive the pictures. Unfortunately for Baird, the BBC decided in 1937 to stop using Baird's system and adopted an all-electronic system instead. Baird was greatly disappointed by the decision and died a bitter man nine years later.

Guglielmo Marconi with his radio transmitter and receiver.

BAIRD showed his television system in public for the first time in January 1926. He transmitted a small and blurred picture of a ventriloquist's dummy. Baird's television system was a mechanical one. Light from the object being televised passed through a rotating disc containing a spiral of holes. The light fell on a photocell that converted the changing light level into an electrical signal. The signal was broadcast to the receiver where it was used to light a bulb. The light from the bulb shone through holes in a rotating disc onto a screen, producing a picture of the object. →

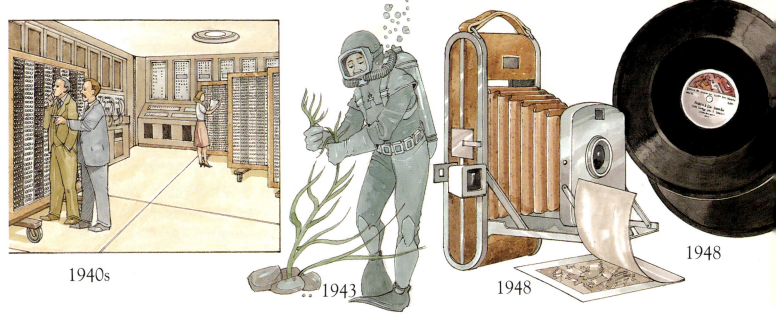

1940s

1943

1948

1948

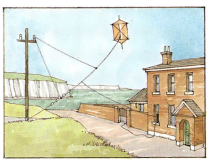

THE FIRST radio signal across the Atlantic was picked up by an aerial carried aloft by a kite. Marconi knew that the weaker the signal, the higher the aerial had to be.

ON 6 AUGUST, 1945, the Americans exploded the first atom bomb on the Japanese city of Hiroshima. Three days later another was exploded on the city of Nagasaki. Shortly afterwards the Japanese surrendered and the Second World War ended.

1940 Russian-born US engineer Igor Sikorsky made the first flight in his helicopter, the VS-300.

1940s The ball-point pen becomes popular. It had been invented in 1938 by the Hungarian artist and journalist Ladislo Biro in Argentina. Visiting a printers, he was impressed by the quick-drying ink they used and decided to make a pen which used it.

1941 English engineer Frank Whittle had the idea for the jet engine in 1928. The first airplane with a Whittle jet engine flew in 1941.

1941 US inventors Lyle Goodhue and William Sullivan develop the aerosol spray can from an earlier idea of the Norwegian scientist Erik Rotheim.

1940s Development of the computer began. The first all-electronic computer, ENIAC (Electronic Numerical Integrator and Computer), was completed in 1946. It could do calculations in an hour that had previously taken a year.

1943 Jacques Cousteau, a French naval officer, invented the aqualung. It had a special valve that let the diver breathe from a tank of compressed air on his or her back.

1948 American inventor Edwin Land produced the instant photographic camera. It could deliver a finished black and white photograph in about a minute.

1948 The 'long-playing' record was introduced by the CBS Columbia Corporation in the USA.

1950

1954

1954

INVENTIONS OLD AND NEW

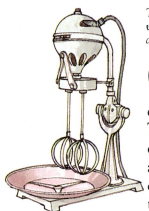

The first electric food mixer was an egg whisk made in 1910 by George Schmidt and Fred Osius of Wisconsin, USA.

Our homes contain many inventions. These inventions can be complicated devices, such as a food mixer. Or they can be so simple that we take them for granted – a plate, for example. In Europe, plates were used by the Romans, but they disappeared when Rome fell in 410. People ate from bowls or put their food on thick slices of bread called trenchers. Plates were re-invented in 1530 for the banquet celebrating the marriage of King Francis of France.

The handkerchief, too, was invented by the Romans. Like the plate, the handkerchief disappeared when Rome fell and did not reappear until the 1400s in Italy. At that time, stylish people carried two handkerchiefs, one for the nose and the other for wiping a sweaty forehead.

The fork is another ancient invention which vanished for a while. Early forks have been dug up from ancient ruined cities in Turkey. No other forks are known until the 11th century when they were used in Italian households to eat fruit which would otherwise stain the fingers.

THE FIRST electric oven was installed in 1889 in the Hotel Bernina near St Moritz, Switzerland. The oven shown was made in the 1920s.

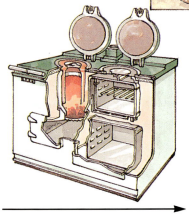

THE blind Swiss scientist Gustav Dalen invented the Aga cooker in 1924.

FROM 1892 heating elements in cookers were cast-iron plates with electric wiring underneath which provided the heat. Since 1926 these have been replaced by elements that can be bent into any shape for ovens and hotplates. This illustration is based on a US advertisement published in the 1950s. At that time eye-level ovens and dishwashers were quite uncommon in Europe.

1954

1957

1957

1958

THE automatic dishwasher was invented by Eugène Daugin in 1885. The first electrically powered dishwasher was made in 1912 but it was not until 1940 that a really successful automatic dishwasher came onto the market. This one dates from the 1950s.

1950 Credit cards were invented in the USA by businessman Ralph Schneider. He launched the Diners' Club, whose members could use plastic cards to pay for meals in restaurants.

1954 British boat designer Christopher Cockerell experimented with a vacuum cleaner blowing air into upturned coffee tins. A year later, he patented the idea of a craft that floats on a cushion of air: the hovercraft.

1954 The first transistor radio, the Regency TR-1, went on sale in the USA. Using transistors instead of the larger thermionic valves meant that the 'tranny' was small enough to be carried about.

1954 US researcher Gregory Pincus invented the contraceptive pill which prevents women getting pregnant. The first large-scale trials were made in Puerto Rico in 1956. It is now used by millions of women throughout the world.

1957 The first man-made satellite, *Sputnik 1*, was launched by the USSR on 4 October, 1957. It stayed in orbit for 92 days and then burnt up in the atmosphere as it fell to earth.

1957 The endoscope is a flexible tube used by surgeons to look inside the human body. In 1957, US doctor Basil Hirschowitz invented the fibrescope, an endoscope that used thin glass fibres to direct light to the surgeon's eye.

1958 Swedish doctor Ake Senning invents the heart pacemaker which stimulates a weak heart with minute electrical shocks.

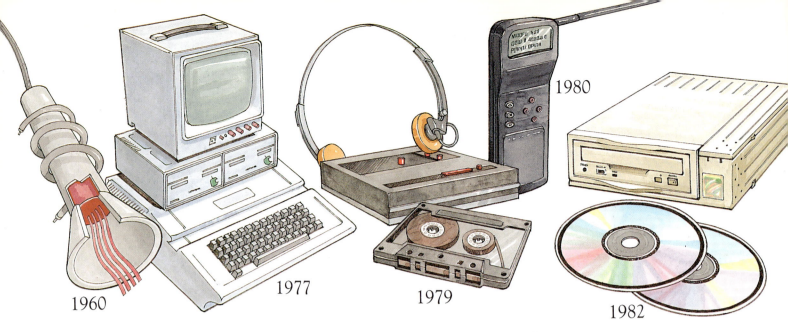

1960

1977

1979

1980

1982

THE COMPUTER AGE

Frenchman Blaise Pascal and his 1642 adding machine. It had eight dials on top to indicate the numbers being added.

The first computers were mechanical devices. They contained cogwheels which turned as numbers were added or subtracted. Frenchman Blaise Pascal made the first mechanical adding machine in 1642 when he was 19 years old.

The first all-electronic computer was built in 1946, by John Mauchly and Presper Eckert in Pennsylvania, USA. ENIAC (Electronic Numerical Integrator and Computer), as it was called, contained 18,000 thermionic valves and occupied a large room.

Modern computers use silicon chips or integrated circuits instead of valves. Integrated circuits contain all the components and connections of an electrical circuit on a small piece of silicon. They are more reliable and use less power than valves. Jack St Clair Kilby, an American scientist, made the first integrated circuit in 1958. In 1971, Ted Hoff, a 34-year-old engineer from Palo Alto in California, USA, invented the microprocessor, a chip that had the main parts of a computer on a single piece of silicon. The microprocessor became the 'brain' of the home computer.

In 1834 English inventor Charles Babbage designed the Analytical Engine, which was a mechanical computer. Unfortunately, the machine was so complicated it was never completed.

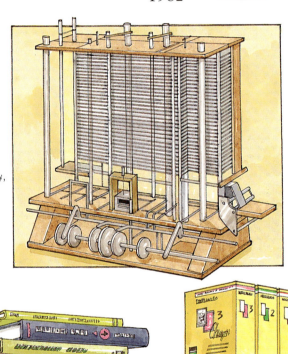

THE computer mouse was invented in 1963 by Douglas Engelbart at the Stanford Research Institute, USA. The mouse is a device that moves a pointer (cursor) on the screen and sends commands to the computer when a button is pressed. In 1983 the Apple Lisa computer combined the mouse and pull-down screen menus, making the computer easy to use. A similar system, Windows, was introduced by Bill Gates of the Microsoft Corporation in 1985.

1982   1989   1989   1991   1995

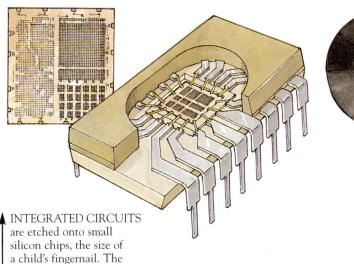

INTEGRATED CIRCUITS are etched onto small silicon chips, the size of a child's fingernail. The chips are then placed in a box with legs, to which the electrical wires are attached.

THE floppy disc used to store data and programs was invented in 1950 by Yoshiro Nakamats of Tokyo, Japan. The discs were first sold by US computer giant IBM in 1970.

1960 US scientist Theodore Maiman built the first laser which produced a thin, powerful beam of light.

1977 The first successful personal computer, the Apple II, was launched in the USA.

1979 The Japanese company Sony introduce the Walkman, a small tape cassette player with light headphones. The tape cassette had been introduced in 1963 by the Dutch firm Philips.

1980 The portable phone became popular in the 1980s. The phones, also called cell phones, are linked by radio to the regular telephone network.

1982 Philips and Sony introduce the compact disc. Only 12 centimetres across, it can hold an hour of sound.

1982 The first combined video camera and recorder, or camcorder, introduced by Sony.

1989 The Japanese company Nintendo introduced the Game Boy, a hand-held video game.

1989 The first broadcasts of high-definition television began in Japan in 1989.

1991 Virtual reality arcade games were introduced in 1991. These games produce the illusion of being in another world.

1995 A high-temperature super-conductor which can be used in brain scanners was invented in the USA.

AMAZING INVENTIONS

In 1853 a customer at the Moon Lake House Hotel at Saratoga Springs, New York, complained that the fried potatoes were cut too thick. The chef, George Crum, did not take the complaint lightly. He hit back by producing a plate of the thinnest potato slices ever seen. When the customer contentedly cruched his way through a second helping, Crum knew he had invented a winner – the potato crisp.

A pedal-powered aircraft was designed by Frenchman J-J Bourcart in 1863. The inventor was certain it would fly, but history does not record if the craft was ever built and tested. One important feature of the craft, according to Monsieur Bourcart, was that the pilot had both hands free during flight.

American inventor William Richardson produced his swimming machine in 1880. The swimmer lay along a central shaft and used hands and feet to turn a propeller. Mr Richardson promised that speeds of up to 10 km/h could be reached.

In 1881 Benjamin Atkins of Cincinnati, USA, invented a finger-supporting device for the piano player. The device was said to improve the player's touch and make playing less tiring.

Mr Traugott Beek of New Jersey, USA, invented an interesting life preserver in 1877. The top part consisted of a floating buoy, with a roof which could be pulled over the user's head in stormy weather. A window allowed a lookout to be kept. The lower part of the life preserver consisted of waterproof trousers and boots, covered in metal bands to protect against sharks. The life preserver held enough food and drink for a month.

The first made-up language was Esperanto, invented by Doctor Lazarus Ludwig Zamenhof in 1878. He made up the language using a twenty-eight letter alphabet, a small number of basic words and a simple set of grammar rules. Zamenhof believed that if all people spoke the same language there would be fewer wars and disputes.

The first solar-powered motor was demonstrated by Abel Pifre in Paris on 6 August, 1882. It consisted of a concave mirror 3.5 metres across which concentrated sunlight on a boiler. When the water boiled, the steam produced powered a steam engine. This, in turn, powered a printing press which worked continuously for $4\frac{1}{2}$ hours, even though the sky was often cloudy.

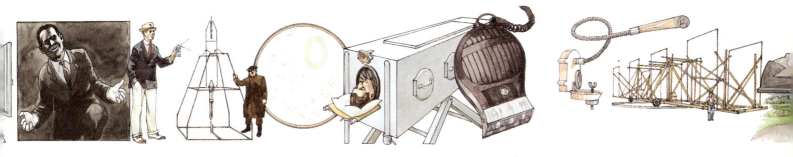

Nathaniel Brown of Kansas, USA, designed the Swing bicycle in 1887. It had two giant wheels 3.5 metres across. A basket hanging from the axle between the wheels held two riders. The basket was made to swing by pulling levers. This made the bicycle roll along. To steer, one rider pulled harder than the other.

The first Meccano set was invented by Frank Hornby of Liverpool, England, in 1900. It was made to amuse his two young sons. The metal strips with holes punched at regular intervals could be bolted together to make almost anything. When he started selling the sets, Hornby chose a rather dull name, Mechanics Made Easy. In 1907 he adopted the name Meccano.

The world's first crossword puzzle was published in the *New York World* on 21 December, 1913. It was composed by Arthur Wyne. The new game quickly became popular. In 1924 the first book of crossword puzzles was published by Simon and Schuster of New York.

The first radio receiver to use a frog's leg was built by Professor Lefeuvre of Rennes, France, in 1921. He connected the frog's leg to the electrical circuits of a radio. When the radio produced an electrical current, the leg twitched and moved a

small pen attached to it. The movements of the pen were recorded on the surface of a revolving drum coated with soot. The purpose of the invention was not entirely clear.

French poet and inventor of the gramophone, Charles Cros, invented a method of signalling to Mars and Venus. The scheme involved the construction of huge mirrors which focused a light beam towards the distant planets. In the 1930s, plans were made to put the scheme into practice, but it proved too expensive.

The game of Monopoly was invented by Charles Darrow of Germantown, Pennsylvania, USA, in 1933. He first sketched the game out on the cloth covering his kitchen table. This was the time of the Depression, when many people were out of work, so Darrow's friends and family enjoyed playing at being wheeler-dealers, buying and selling properties with pretend money.

The Bloggoscope was invented in the 1950s by Bill Bloggs to help make maps from aerial photographs. The Bloggoscope was a 'tilt finder'; it found the angle that the aerial camera was tilted in relation to the ground when a photograph was taken. This information was vital if an accurate map was to be drawn from the photograph.

GLOSSARY

Alternating current Electric current that flows first in one direction and then in the opposite direction, reversing its direction many times a second.

Bearing Part of a machine that allows a moving part to turn or slide more easily.

Condenser Device in which a gas condenses, or changes, to a liquid. In a steam engine, the condenser changes steam to water.

Cylinder The tube inside an engine in which the fuel burns and the piston moves.

Direct current Electric current that flows only in one direction. The electricity produced by a battery is direct current.

Electromagnetic wave Wave of electric and magnetic force that carries energy through empty space. Light, microwaves, ultraviolet rays, infra-red rays, X-rays, radio waves are all electromagnetic waves.

Filament Very thin wire in an electric light bulb which glows white hot when an electric current is passed through it.

Four-stroke engine Type of internal combustion engine in which the power is produced in four strokes, or movements of the piston. Automobile engines are usually four-stroke engines.

Gear wheel Toothed wheel that transmits the turning movement of one shaft to a second shaft. Gears are used in automobiles to transmit the power of the engine to the wheels.

Generator Machine that produces electricity from moving magnets.

Integrated circuit Complete electronic circuit built on a small piece of silicon, called a chip. Integrated circuits are used in computers and other electronic devices.

Internal combustion engine Petrol or diesel fuel-burning engine, as used in most motor vehicles.

Jet engine Type of gas-turbine engine used in most large aircraft. A stream of hot gases is expelled from the rear of the engine, driving the aircraft forward.

Lever Simple machine used for lifting heavy weights. It consists of a strong bar that rests on a pivot, like a see-saw.

Microprocessor Integrated circuit that contains most of the parts of a computer on a small piece of silicon. They are used in small computers and in many devices around the home.

Piston The part of an engine which moves inside the cylinder.

Pulley Simple machine for lifting weights. It consists of a wheel with a groove around the rim in which a rope runs.

Receiver Device that picks up, or receives, a signal or message.

Satellite Object that circles around another much larger body. The moon is a satellite of the earth.

Silicon Chemical element which occurs widely in minerals and in sand. It can be a brown powder or a brown crystal.

Silicon chip Another name for an integrated circuit; a small piece of silicon with electronic circuits etched onto its surface. Silicon chips are used in computers and other electronic devices.

Superconductor Substance which loses its electrical resistance at very low temperatures.

Thermionic valve Electronic component consisting of a glass bulb with terminals attached, used to control electric currents.

Transformer Device used to change the voltage of an alternating electric current. Transformers are used in electric power stations.

Transistor Electronic device which controls the flow of an electric current. It is used in radios, television sets and computers.

Transmitter Device which sends or transmits a signal or message.

Two-stroke engine Internal combustion engine that produces its power with two strokes of the piston.

Vacuum Empty space, that contains no air or other material.

Virtual reality The illusion of an artificial world produced by a computer. Some computer games use virtual reality.

Voltage The electrical pressure in a circuit, which drives the electrical current around the circuit.

INDEX